Mines and Geology
of the Randsburg Area

Randsburg from the north, circa 1897. The high point on the skyline is Government Peak, and the site of the newly discovered Yellow Aster Mine is in the canyon to the right of the peak. (John W. Robinson Collection.)

Mines and Geology of the Randsburg Area

An Historical Gem of the Mojave Desert

by

D.D. Trent, Geologist and Emeritus Professor
Department of Physical Sciences and Engineering
Citrus College, Glendora, CA 91741

San Diego Geological Society

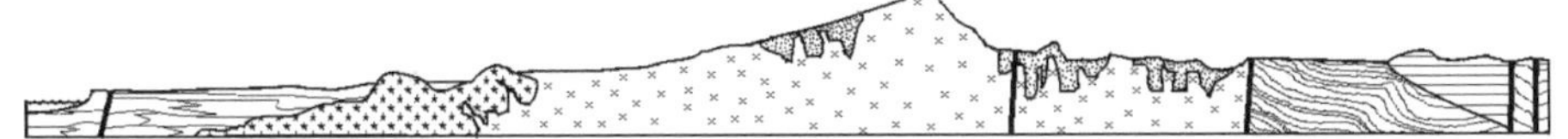

Distributed by Sunbelt Publications
Chula Vista, California
www.sunbeltpublications.com

Mines and Geology of the Randsburg Area:
An Historical Gem of the Mojave Desert

Copyright © 2006 by D.D. Trent
Published by the San Diego Geological Society, Inc.
3130 North Evergreen Street, San Diego, CA 92110
www.sandiegogeologicalsociety.org

Edited with design and composition by W.G. Hample & Associates

Available through:
Sunbelt Publications, Inc.
664 Marsat Ct., Suite A, Chula Vista, CA 91911
(619) 258-4911
www.sunbeltpublications.com

27 26 25 24   9 8 7 6

ISBN 13: 978-0-916251-77-2
Library of Congress Cataloging-in-Publication Data:

Mines and geology of the Randsburg area:
an historical gem of the Mojave Desert /
D.D. Trent. −1st ed.
        p. cm.
        Includes bibliographical references and index
        ISBN 0-916251-77-2
1. Mines and mineral resources- California- Mojave Desert
2. Guidebooks and travel- California- Mojave Desert
3. Geology- California- Mojave Desert
I. Trent, D.D.
        TN24.C2   2006
        553'.   dc22

Cover photo by D.D. Trent.
Photos and illustrations by respective authors as noted and used with permission.

Contents

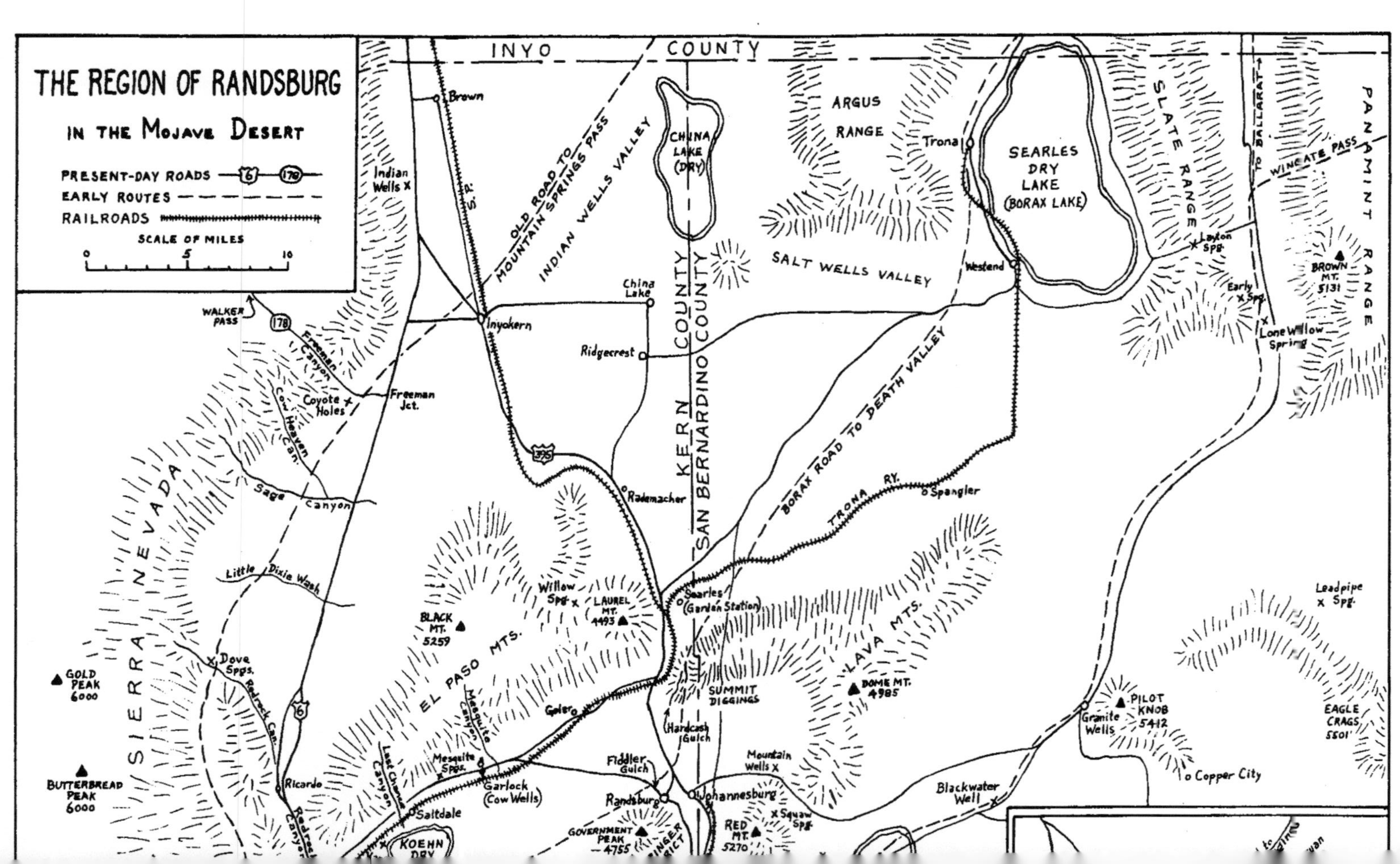

THE REGION OF RANDSBURG
IN THE MOJAVE DESERT
PRESENT-DAY ROADS —(6)—(178)
EARLY ROUTES
RAILROADS
SCALE OF MILES
0 5 10

INYO COUNTY
KERN COUNTY
SAN BERNARDINO COUNTY
Brown
OLD ROAD TO MOUNTAIN SPRINGS PASS
INDIAN WELLS VALLEY
CHINA LAKE (DRY)
ARGUS RANGE
Trona
SEARLES DRY LAKE (BORAX LAKE)
SLATE RANGE
TO BALLARAT
WINGATE PASS
PANAMINT RANGE
Indian Wells X
WALKER PASS
Layton Spg.
BROWN MT. 5131
Early X Spg.
X Lone Willow Spring
China Lake
SALT WELLS VALLEY
Westend
Ridgecrest
Inyokern
Freeman Jct.
Coyote Holes
Freeman Canyon
Cow Heaven Can.
Sage Canyon
Rademacher
BORAX ROAD TO DEATH VALLEY
TRONA RY.
Spangler
SIERRA NEVADA
Little Dixie Wash
Leadpipe X Spg.
Willow Spg. X
LAUREL MT. 4493
BLACK MT. 5259
EL PASO MTS.
Searles (Garden Station)
LAVA MTS.
DOME MT. 4985
X Dove Spgs.
GOLD PEAK 6000
Goler
SUMMIT DIGGINGS
Hardcash Gulch
Mountain Wells X
PILOT KNOB 5412
Granite Wells
EAGLE CRAGS 5501
o Copper City
BUTTERBREAD PEAK 6000
Ricardo
Mesquite Canyon
Mesquite Spg. X
Garlock (Cow Wells)
Fiddler Gulch
Randsburg
Johannesburg
X Squaw Spg.
Blackwater Well X
Saltdale
Last Chance Canyon
Redrock Can.
KOEHN DRY
GOVERNMENT PEAK 4755
RED MT. 5270
MINING DISTRICT

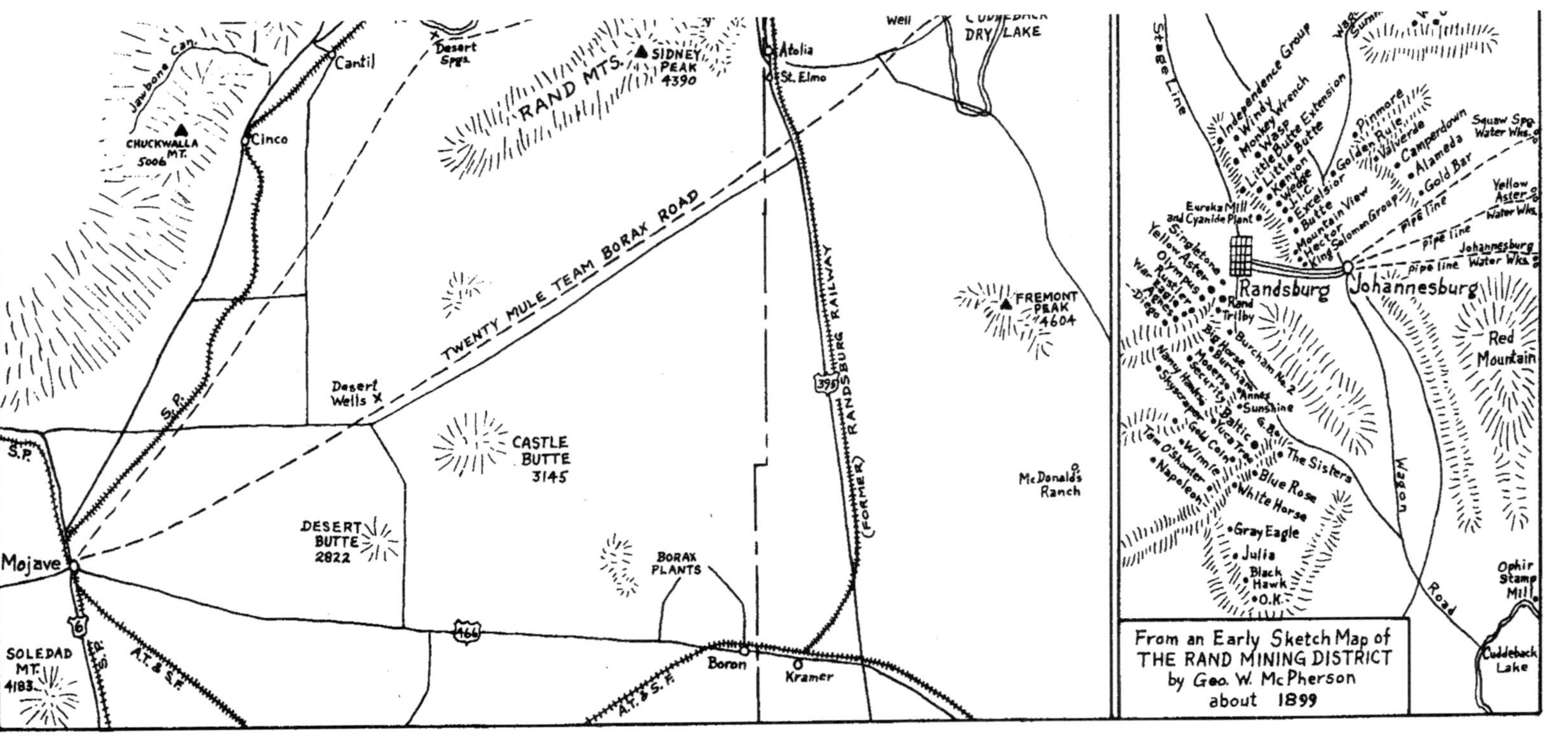

Map of the Randsburg, California, region.

From Wynn, 1963, Desert Bonanza: The Story of Early Randsburg, Mojave Desert Mining Camp.

(Courtesy of Arthur H. Clark Company.)

Preface and Acknowledgments

In the history of the mining camps of the Old West, the names of Cripple Creek, Bodie, Tombstone, Leadville, Tonopah, and Virginia City come to mind. The stories of these and many other Western mining camps have been told often and well. But it is lesser-known Randsburg and the Rand, and its nearby Red Mountain and Atolia Mining Districts, that is truly unique among mining camps. Collectively, the area has had five lives in contrast to the single boom-and-bust cycle so common in the West's mining scene.

This booklet is intended to tell a different story than is usually told in the many books on the history of Western mines, the story of the geology and how the miners got the gold, silver, and tungsten. In order to do this, I'm indebted to many geologists who have studied and written on the area, among whom are Bill Clark, Tom Gay, George Hazenbush, Carlton Hulin, Paul Morton, Richard Stewart, Benny Troxel, and Loren Wright whose important geologic and mining reports on San Bernardino County and Kern County were published by the California Division of Mines and Geology (now the California Geological Survey). Benny Troxel also read the manuscript and made several substantial suggestions. Andrew Barth of Indiana-Purdue University and Carl Jacobson of Iowa State University provided isotopic age dates of rocks in the Randsburg region. Greg Brasel, geologist with Rand Mining Company, provided information on the modern mining and extraction processes of the Yellow Aster and Baltic Projects and also arranged visits to the mine by my geology students. Also, my thanks go to Michelle Nielsen, archivist at the San Bernardino County Museum, for her helpful assistance in locating important photographs, and to John W. Robinson who loaned me out-of-print books and encouraged me to rummage through his collection of historical photographs.

Thanks, too, must go to Margaret Eggers and Lowell Lindsay of the San Diego Association of Geologists who felt my scratchings worthy of publishing, and to Bill Hample, graphic designer and editor who massaged and organized my manuscript and illustrations. Any errors of commission or omission are, of course, my responsibility.

D.D. Trent
Claremont, California

Introduction

*What Cripple Creek is to Colorado
so will the Rand be to Southern California—the most extensive and richest mining section in California.*
—Anonymous, *The History of the
Rand Mining District*, 1898

Except for copper mining in Arizona and Montana, the great mining booms of the Western states had faded by the 1890s. In Nevada and California, the low price of silver and the depletion of the rich gold and silver ore bodies brought mining to a virtual standstill. Furthermore, the financial collapse of 1893 clearly had made the Gay Nineties something less than gay.

But then along came Randsburg which rejuvenated California mining.[1] It is doubtful that any mining region in the United States has been more fortunate than the region of the Rand (also called the Randsburg) Mining District and its adjoining Atolia and Red Mountain Mining Districts. The region has gone through five mining booms in contrast to the single boom-and-bust cycle experienced by so many Western mining camps.

Initially, the Rand Mining District was a gold camp, but the escalating costs of mining during World War I, "The Great War" (1914-1918), forced closure of most gold mining operations in the district. However, the war created a need for tungsten, and that prompted the feverish development of tungsten properties at Atolia, four miles southeast of the gold mines. After signing the Armistice in 1918, the high costs of gold mining were still prohibitive, the market for tungsten collapsed, and mining in the district almost ended. But Congress' passage of the Pittman Act in 1919—guaranteeing the price of domestic silver at $1.00 an ounce—breathed new life into the district, prompting the discovery of silver deposits, and the district boomed for a third time

---

[1] Coincident with the gold discoveries in the western Mojave Desert were similar discoveries in Alaska and the Yukon that resulted in 100,000 gold seekers becoming the Klondikers of '98. To some extent, these gold rushes helped relieve the burden of the nation's financial crash of the 1890s.

(Hulin, 1925, p. 15). But when the Pittman Act expired in 1923, the price of silver dropped to 65 cents, and most silver mining collapsed again, except for the California Rand Mine at Red Mountain, which continued mining silver profitably.

When the United States joined the allied nations in 1941 to become a combatant in World War II, the urgent need for tungsten (used for hardening steel and as filaments for incandescent light bulbs), demanded by the war effort, stimulated renewed mining in the district for the fourth time. This continued after the war until 1956, when the need for tungsten was satisfied, and the district again fell on hard times. Then, following the lifting of federal restrictions on gold prices in 1968, the Rand District took on a new life for a fifth time (Rand Mining Company, 2001, p. 2).

Looking east along Butte Street, Randsburg, circa 1898. Little Butte and King Solomon Mines are located near the skyline on the right side of the photograph. (John W. Robinson collection.)

Chapter 1
Gold at Randsburg

I wanted the gold, and I sought it;
I scrabbled and mucked like a slave.
Was it famine or scurvy—I fought it;
I hurled my youth into a grave…
There's gold, and it's haunting and haunting;
It's luring me on as of old;
Yet it isn't the gold that I'm wanting
So much as just finding the gold.

—Robert W. Service, *The Spell of the Yukon*, 1907

Initially prospected in the 1860s, it was not until 1893 that placer gold was discovered in Goler Wash in the El Paso Mountains, eight miles west of Randsburg, that any serious effort at mining began in the western Mojave Desert. The efforts of obtaining gold from the surface gravels were limited because of the shortage of water in the region. This resulted in numerous dry washing[2] camps springing up in this part of the Mojave. The two areas with economic gravels in the Randsburg region were the valley north of and adjacent to the Rand Mountains, and the gulches of the Stringer District, a part of the Rand District, about two miles south of Randsburg, on the southern and eastern sides of Government Peak. The name of the district originates from the occurrence of gold-bearing ores along fault zones in "stringers" rather than in major lodes or concentrated ore bodies. The Stringer District contained some of the more productive mines of the Rand District, including the Gold Coin, Baltic, Hawkeye, and others.

---

[2] Dry washing is a gold-separating process used in arid regions where there is little or no water. Gold-bearing dirt is placed in a hopper at the top of a dry-washer through which air, from a bellows pumped by hand or by a gasoline engine, winnows away the lighter mineral particles, allowing the heavier, more valuable minerals to be caught behind riffles.

In 1895 the Olympus Mine (later to be named the Yellow Aster), was discovered by John Singleton, C.A. Burnham, and Fred Mooers (Hess, 1909, p. 31). The district took its name from the Rand District of South Africa, and the local town that serviced the mines became Randsburg (Clark, 1992, p. 164). Once the news became known, the region was teeming with prospectors who rushed to the area in the search for new mines. A rash of discoveries soon followed and another gold rush was underway. The prospectors and miners gave picturesque names to their discoveries, names such as Minnehaha, Gold Coin, Gold Bullion, Silver Bell, King Solomon, Monkey Wrench, and Treasure Hill. And what prompted those nineteenth-century miners to name the district in honor of South Africa's Rand and use such picturesque names for their mines? It's because in those days it was important to select imposing names for mining districts and mines in order to attract investors or purchasers. Mining historian Otis Young points out that humor, too, could be expressed in naming mines.

> A case in point is that of Amanda Reed Shoemaker…, who as a little girl in the Black Hills had a promising lode claim, the Little Allie, recorded in her honor by a friendly neighbor. The neighbor's wife reportedly grew resentful and said to him, "You've named claims for your favorite politicians, for your old girl friends, and even one for your pack mule. Now, I've looked it up, and I'm told you can amend the location to change the name, and I demand that you do just that and name this one for me!" Her husband gave in and recorded the amended location as the Holy Terror, which name it still bears. (Young, 1970, p. 32.)

Randsburg only had 13 buildings in 1895; most were made of canvas, with a few sporting wooden false fronts. It was little more than a tent camp, but by early 1897, Randsburg boasted some 300 buildings and tents. With a population of about 2,500, it became one of the great boom towns of the West (Vredenburgh et al., 1981, p. 195; Nadeau, 1965, p. 285). It even had a rival town, Johannesburg, two miles to the northeast. Johannesburg at its height had two general stores, boarding houses, livery stables, a billiard hall, music hall, lunch counter, barber shop, real estate office, two laundries, two lumber yards, a telephone exchange with Randsburg, and a railway station, at the northern terminus of the Randsburg Railway, which made connections with the Atchison,

The Baltic Mine, ten-stamp mill, miners' cabins in the background, and placer miners working gold-bearing gravels in the foreground, on the northern end of the Stringer District, probably 1898. The gravels would be reworked for tungsten ore during World War I. In the 1990s, this became the site of the Rand Mining Company's Baltic Project. (Wynn collection courtesy of Arthur H. Clark Company.)

Placer miners working gold-bearing gravels with dry washers in the Stringer District, about two miles south of Randsburg, 1898. (Wynn collection courtesy of Arthur H. Clark Company.)

Topeka, and Santa Fe at Kramer Junction. In 1900, Johannesburg was able to boast of a golf course and a golf club with 13 members including seven women (Wynn, 1963, p. 135; Myrick, 1963, p. 796).

Three shafts and approximately fifteen miles of underground workings in fourteen levels were developed in the Yellow Aster. The shallower workings proved to be the most profitable, with values of some of the shallow ore reported to have been $1,000 per ton. In the first years of operation, the ore was hauled eight miles to Garlock by freight teams for processing in small stamp mills. Later the ore was sent by wagons to a custom mill at Barstow.

It was the prospect of substantial revenue from shipping Yellow Aster ore to the mill at Barstow, as well as income from freight and passenger service, which attracted a number of investors interested in backing a railroad; the result was the construction of the Randsburg Railway. The 28.5-mile branch line from Kramer Junction to Johannesburg was constructed in 1897. The first train rolled into Johannesburg on December 27, 1897, and scheduled operation with two round trips daily began in January 1898.[3] Although it was only one additional mile to Randsburg, the railroad never reached its namesake because of steep grades and other considerations (Myrick, 1963, pp. 794-798).

The substantial revenue for the railroad for ore traffic from the Yellow Aster to the mill at Barstow only lasted a short time. The Yellow Aster built its own 30-stamp mill with amalgamation tables in 1898, and a 100-stamp mill was added in 1899, by which time the Yellow Aster operated one of the largest stamp mills in California. Ore was hauled the short distance from the mine to the mill by a narrow-gauge railroad (Troxel and Morton, 1962, p. 128; Myrick, 1963, p. 798). By the early 1900s, Singleton, Burnham, and Mooers, the discoverers of the Yellow Aster, had achieved financial success, thus making the Rand one of the few mining districts to actually enrich the discoverers (Nadeau, 1965, p. 284).

Eventually, many of the early workings disappeared in a glory hole that in the later years of operation served as the haulage way for

---

[3] The Randsburg Railway was acquired by the AT&SF in 1903. A decline in mining and lack of business resulted in abandonment of the branch line in 1933, with the rails removed in 1934.

Randsburg Railway locomotive No. 1 and train at the Johannesburg station as pictured in the Randsburg Miner, *November 17, 1900. (Courtesy of Phil Serpico.)*

bringing ore to the surface (Hulin, 1925, p. 122). By 1962 the Yellow Aster, the major producer of the district, was credited with an output of $12 million, the gold, alloyed with silver, averaging about 750 fine (Troxel and Morton, 1962, p. 128). Other major gold producers in the district and their output are the Butte, $2,000,000; Sunshine, $1,000,000; Blackhawk, $700,000; and Operator Divide, Big Gold, Buckboard, and King Solomon, each producing $500,000 or more. At least six other mines yielded more than $250,000 in gold (Tucker et al., 1949, pp. 214-216; Clark, 1992, p. 167: Troxel and Morton, 1962, p. 50).

Most of the district's ore is hosted in the Rand Schist,[4] the predominant rock unit of the district, and in the Randsburg granodiorite

---

[4] The Rand Schist has a complex geologic history. It has long been considered to be of Precambrian age, and it shares many geological (lithologic) characteristics with the Orocopia Schist of southeastern California, and the Pelona Schist of the San Gabriel Mountains. It is now recognized that each of these schist units have different geologic histories, and the minimum $^{40}Ar/^{39}Ar$ isotopic age for metamorphism of the Rand Schist is in the range of 73 to 68 million years (Late Cretaceous) (Grove et al., 2003, p. 381; Jacobson, pers. communication, 27 Aug 2005; Jacobson et al., in press).

and the Atolia granodiorite. The Randsburg granodiorite is a Miocene intrusive rock, yielding an isotopic age of 19 to 20 million years, and the Atolia granodiorite is Late Cretaceous, dated at 79 million years (Barth et al., 2003). Other rocks of Tertiary age in the district include dioritic and andesitic intrusive rock and rhyolite dikes, and the age of mineralization is considered to be Tertiary (Brasel, 2001).

In general, the ore is localized in hydrothermally mineralized quartz veins that formed in fractures in fault breccia. The gold and silver occurs in many small, closely spaced vein systems where hydrothermal activity was probably focused along structurally prepared sites. Typical minerals accompanying the gold include quartz, dolomite, arsenopyrite, siderite, mariposite (chromium mica), pyrite, galena, silver, and scheelite (Emmons, 1937, p. 126). Significant gold mineralization also occurs along the margins of some of the Tertiary dioritic and rhyolitic dikes in the area (Troxel and Morton, 1962).

Nearly all of the mines are developed in a gossan, a chemically decomposed rock which is colored rusty brown due to the oxidation of pyrite and other metallic sulfide minerals. Limits of gossan mineralization are indicated on the accompanying geologic map.

Clark (1992) reported:

> The ore bodies most commonly occur in the vein footwalls, usually at or near vein intersections or in sheared and brecciated zones. The ore consists of iron oxide-stained brecciated and silicified rock containing native gold in fine grains and varying amounts of sulfides. The sulfides increase at depth, but the gold values decrease. Most of the mining stopped where unoxidized sulfides were found in the veins, and the maximum depth of development is 600 feet. Milling ore contains from 1/7 to 1/4 ounce of gold per ton. The high-grade ore nearly always occurs in pockets near the surface. Some placer deposits have been mined but the shortage of water limited most placering to dry washing, mainly at Stringer, and in the Rand Mountains [south and southwest] of Randsburg (pp. 166-167).

The district's gold production was substantial during a minor boom beginning in the 1930s when the price of gold rose from $20.67 an ounce to $35. But the boom ended in 1942 when gold mining was curtailed for the duration of World War II by the War Production Board's

Limitation Order L-208, which shut down marginal gold producers in order to free miners for military service or to work in mines producing strategic minerals needed for the war effort. By 1990 the total gold output of the Rand District was estimated to exceed $20 million (Clark, 1992, p. 164).

The Yellow Aster headframe, hoist house, and glory hole, 1930. The Yellow Aster was developed by three shafts and about 15 miles of underground workings distributed among 14 levels. The mine operated continuously from 1895 to 1918 when it shut down. It reopened in 1921 and continued operating until 1939. Since then it was worked intermittently by lessees until mining rights to the property were acquired by the Rand Mining Company in 1984. By 1962 the mine was the principal source of gold in Kern County, with an official output of approximately $12 million; the gold, alloyed with silver, averaged about 750 fine. The Yellow Aster accounts for about one-quarter of Kern County's total gold output. (Used by permission of the Department of Conservation, California Geological Survey.)

Yellow Aster Mine, 30-stamp mill, ore bin, and shop complex, 1896, one year after the mine's discovery. (San Bernardino County Museum, A2623-1.1716.)

The Butte Mine at the east end of Butte Street, Randsburg. Mill tailings are in the immediate foreground, February 1977. Total production from the Butte was nearly $2,000,000 in gold and silver, with a few units of tungsten. The average ore grade has been 0.75 ounces of gold per ton. Vein material assaying less than 0.25 ounces of gold per ton was stored on mine dumps and was screened with the fines milled in the 1950s. It was estimated in 1962, when gold was valued at $35 an ounce, that the fines in the dumps contained about $5 of gold per ton. (Photo by D.D. Trent.)

Central area of the Minnehaha Mine, about 1.25 miles southwest of Randsburg, on the flank of Government Peak, 1958. The mine had two principal shafts, shown here along with the portal to an adit, and four or five shafts of lesser importance. Several gold-bearing veins extend across the hill. The mine was discovered in 1895, and the principal periods of activity were 1895-1923 and 1931-1941. Gold values recovered are estimated at $100,000. (San Bernardino County Museum, A2623-1.1733.)

King Solomon Mine in 1895 or early 1896. The mine's greatest output was between 1919 and 1942; it eventually produced more than $500,000 in gold. During this period, gold was priced at $20.67 until 1933 when the price was raised to $35 an ounce. (San Bernardino County Museum.)

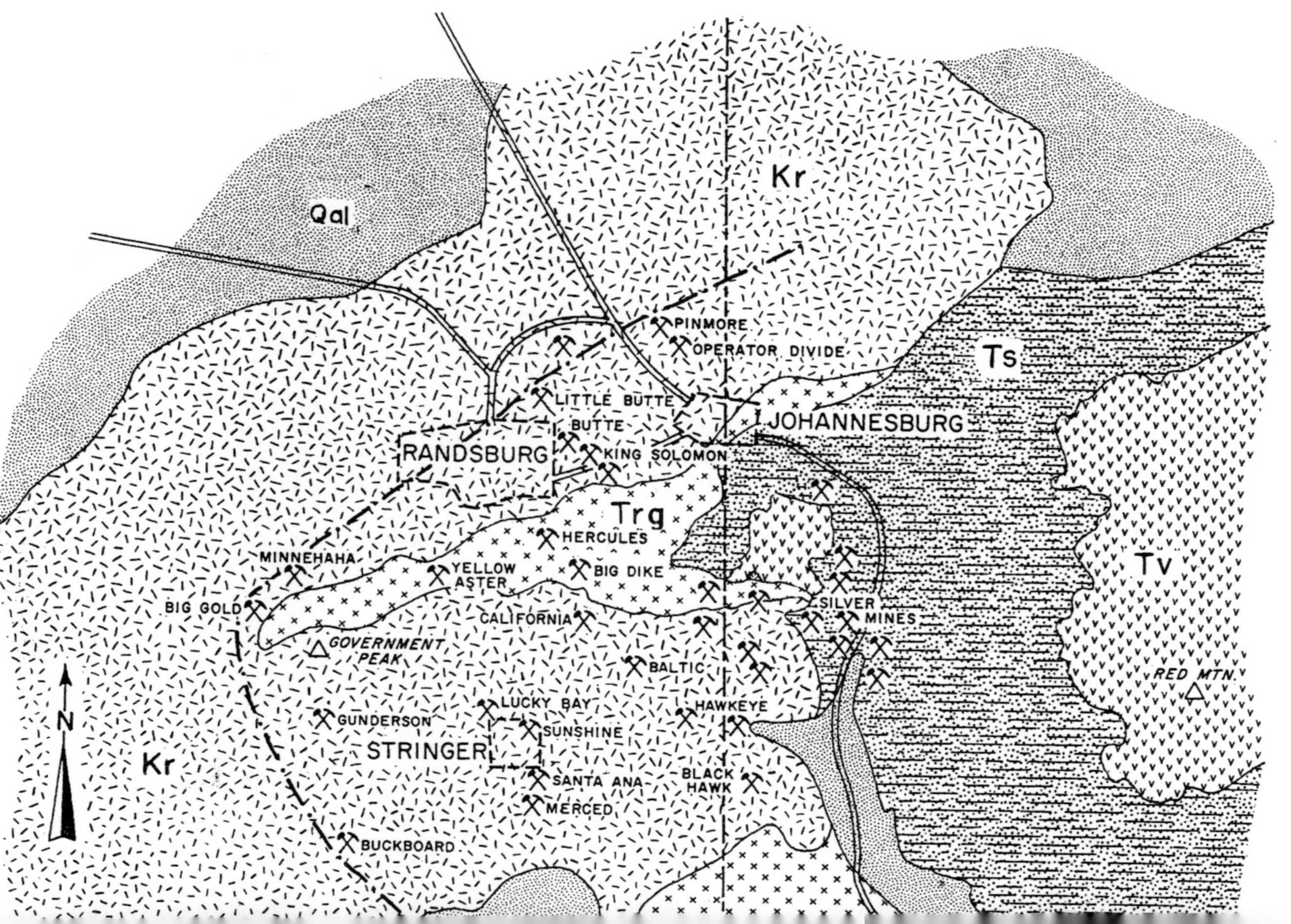

Qal
Kr
Ts
Tv
Trg
PINMORE
OPERATOR DIVIDE
LITTLE BUTTE
BUTTE
JOHANNESBURG
RANDSBURG
KING SOLOMON
HERCULES
BIG DIKE
YELLOW ASTER
MINNEHAHA
BIG GOLD
CALIFORNIA
GOVERNMENT PEAK
SILVER MINES
BALTIC
HAWKEYE
RED MTN.
LUCKY BAY
GUNDERSON
SUNSHINE
STRINGER
SANTA ANA
BLACK HAWK
MERCED
BUCKBOARD
N
Kr
10

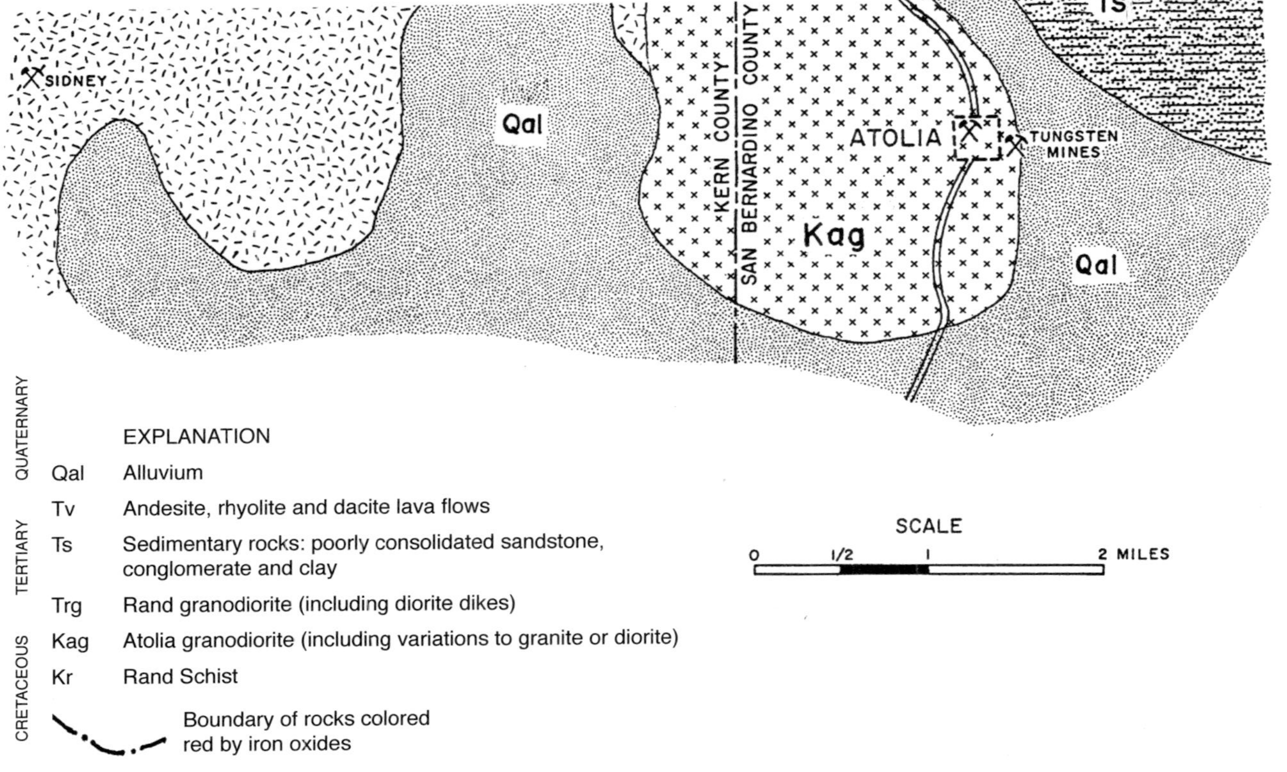

Generalized geologic map of the Rand, Red Mountain, Atolia, and Stringer Districts, showing the locations of the major mines. (Modified from Clark, 1970; Hulin, 1925, Plate 1; [Used by permission of the Department of Conservation, California Geological Survey] rock ages from Barth et al., 2003, and Jacobson, et al., in press.)

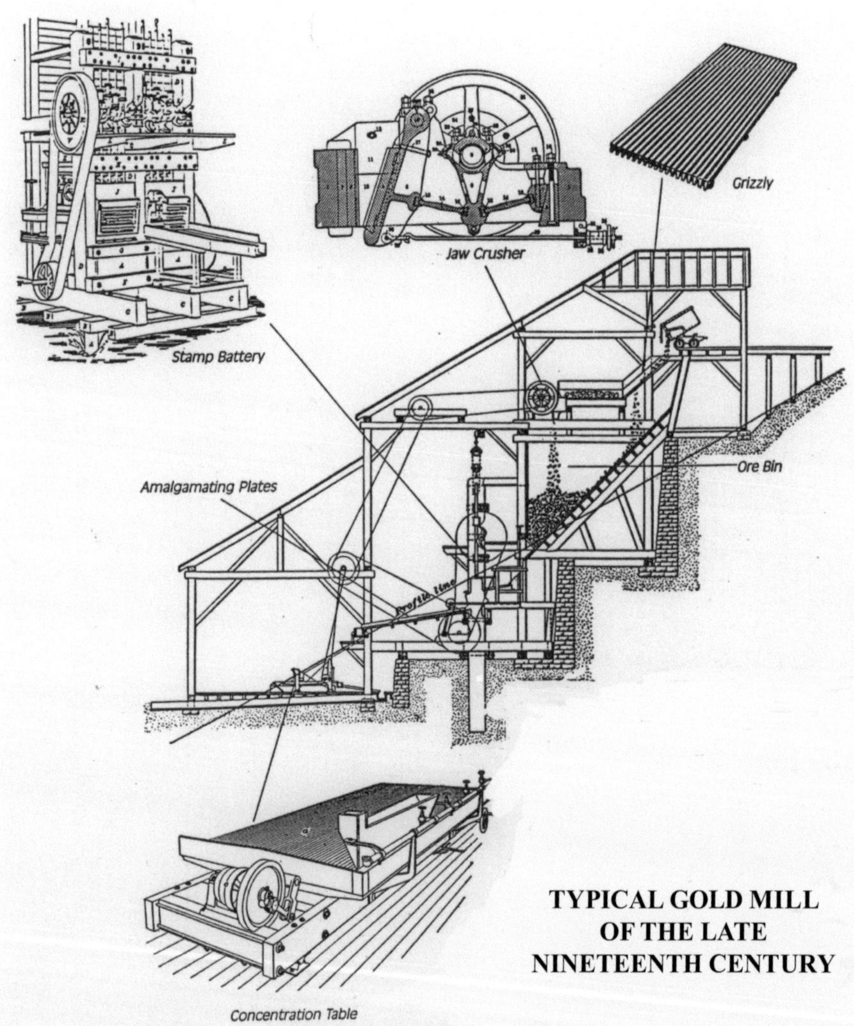

**TYPICAL GOLD MILL
OF THE LATE
NINETEENTH CENTURY**

In this illustration of a gold mill, the machinery and steps in ore processing are similar to that at the Yellow Aster. (1) Ore was dumped onto a grizzly which separated the larger pieces that were routed to the jaw crusher and reduced to two-inch size. (2) Broken ore was stored in the ore bin. (3) Ore was routed to the mortar in the stamp battery, mixed with water and mercury, and crushed to sand sizes. (4) Pulp from the stamp mill flowed across the amalgamation table, a mercury-coated copper plate; the amalgam was scraped off, retorted to recover the mercury for reuse, leaving a spongy gold matte. (5) The residual pulp was routed to a concentrating table where the remaining gold was separated by riffles and gravity sorting; this step was vital with ores containing gold-bearing metallic sulfides, as sulfides will not alloy with mercury. (6) The tailings, slimes leaving the concentrator, were routed to the tailings dump. (7) The gold matte was sent to a smelter or the mint. (Modified from Sagsetter and Sagsetter, 1998, *The Mining Camps Speak*.)

Chapter 2
Ore Dressing in the Early Years (or)
How They Got the Gold and Silver

…for since Nature usually creates metals in an impure state, mixed with earth, stones and solidified juices, it is necessary to separate most of these impurities from the ores as far as can be before they are smelted.

—Georgius Agricola, *De Re Metallica*, 1556,
translation by Herbert Clark Hoover
and Lou Henry Hoover, London:
The Mining Magazine, 1912

Mining the ore is the first step in recovering gold, or any metals for that matter. The next step is to concentrate the desired metals in a mill. By 1909 the Rand District had seven mills with a total of 164 stamps (Hess, 1909, p. 32). The concentration process of gold-quartz mills of the early twentieth century was rather simple. The ore was sent to a grizzly, a set of parallel steel rails, where anything too large went to a jaw crusher where it was reduced to about one inch. Next it was fed into a stamp mill[5] where it was pulverized to a powder and mixed with water to produce a slurry called pulp. The pulp washed across amalgamating apron plates, copper plates coated with liquid metal mercury, where gold and silver alloyed with the mercury to form an amalgam.

The amalgam was scraped from the apron plates and placed in a retort which was then heated. The heated mercury boiled off the amalgam to be condensed in a cooling coil, collected, and reused. The precious metals were left behind as a concentrate, a spongy-looking blob in the retort vessel. Commonly, the crudely refined gold was shipped to a smelter for additional refining to remove the metallic impurities and

---

[5] A vintage five-stamp mill from the Baltic Mine is displayed in Randsburg's city park on Butte Street.

Yellow Aster Mine. Headframe, ore bins, and hoist house in lower left; the 100-stamp mill in upper left; Randsburg in the distance. Date unknown. (San Bernardino County Museum, A2623.1.1738.)

Mill workers during cleanup of the amalgamation tables in the Yellow Aster 100-stamp mill, Randsburg, 1912. (Used by permission of the Department of Conservation, California Geological Survey.)

separate the gold from the silver, but in some cases the final refining was done in a smelter at the mill site (Eggers and Trent, 2004, pp. 54-57). The residue of gangue minerals, called tails or tailings, spilled off the lower end of the mercury table, to be carried away to a disposal heap or pond.

This process might recover 70 to 80 percent of the precious metals. However, if sulfides were present in the ore body, gold and silver recovery was even less, as sulfides will not amalgamate with mercury. Commonly, in order to recover the sulfide values, the tailings were routed to a concentrating or jerking table, a gently sloping deck to which is attached a parallel set of narrow strips of wood, called riffles. Concentrating tables function on the principle that fine-grained mineral particles are sorted by density if kept mobile in water and agitated slightly. The riffles concentrate the denser ore mineral grains, such as scheelite or gold, and the shaking of the table gradually moves the gangue, quartz and other unwanted mineral grains, to one side of the table, where it falls in a trough that leads to a tailings disposal site.

Experiments by mill operators in California, Nevada, and Colorado to improve recovery of gold and silver began in the middle nineteenth century. By the early 1890s, great strides in the technology of ore dressing had been made. For example, the introduction of cyanide to treat precious metal ores increased the recovery to 90 percent or more. Additional improvements in milling procedures continued into the twentieth century. The mill circuit at the California Rand Mine of the early 1920s illustrates how far milling and ore dressing for precious metals had progressed in just a few decades. By this time the mill of the California Rand had a daily capacity of 400 tons of ore. The mill circuit consisted of primary crushers, two ball mills, two tube (pebble) mills, two classifiers, flotation cells, and thickeners. The mill flow chart for the California Rand, as it functioned in 1923, is shown in the illustration on page 16, and the following is an explanation of the process.

Ball mills rotate, causing their grinding media (steel balls) to roll, cascade, and fall, with the effect of these motions crushing the ore to fine particles. Tube mills are similar, but they are long rotating cylinders, with lengths that are about twice that of the diameter, that use chert pebbles for the grinding medium. Classifiers are machines that separate sand-sized grains from the slimes in order to treat each of

FLOW SHEET
CALIFORNIA RAND SILVER, INC.
CAPACITY 400 TONS PER 24 HOURS

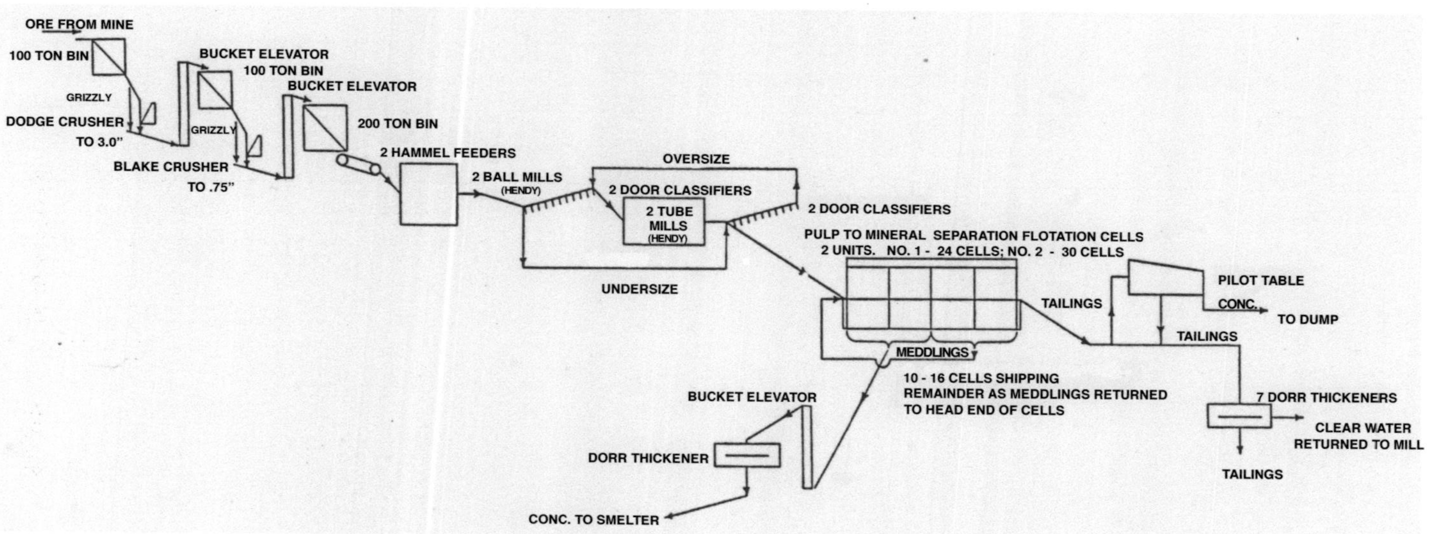

Flow sheet for the mill circuit at the California Rand Silver Mine, Randsburg, California, 1923. (Hulin, 1925.)

these size fractions separately. Flotation cells are mechanical devices for separating particles of different minerals from one another in a water bath, due to differences in their wetability and their reaction to the surface tension of liquids. Metallic sulfides have a tendency to stick to greases or oils, whereas particles of silicate minerals, such as quartz, feldspar, and the micas have no such inclination. Flotation begins by adding a small amount of light oil in the ball mill, then adding it to the water bath in the flotation cells, which also contains an even smaller amount of acid (or another reagent). The acid etches the mineral grains, giving them a clean surface to which the oil can adhere tightly. Furthermore, the oil initiates the formation of a bubbly froth with a high surface tension that is created by a steady stream of low-pressure air

Ball mill similar to those used in the 1920s at the California Rand. Coarse fragments of ore and steel balls, three to five inches in diameter, are fed into these revolving mills. The tumbling and grinding action of the ore and balls pulverizes the ore to a powder. Water is added to allow the powdered ore to flow as a slurry and to prevent dust. (Allan B. Marquand Collection, San Juan County Historical Society, Colorado.)

entering the bath from an orifice at the bottom of the cell. The air creates bubbles that are enclosed in a monomolecular film of oil. The low-wetability metallic sulfide grains adhere to oil and the bubbles that cause the desired sulfides to rise in the froth to the top of the bath. Here, encouraged by revolving paddles, the froth spills over the side of the flotation cell into a trough that carries away the muddy but valuable slurry. From here it is dried and sacked and is known as concentrate. Meanwhile, the undesired mineral grains sink due to their high wetability, to be washed away by the currents in the bath and drawn off to eventual disposal on the tailings pond (Trent, 2005, p. 366).

A bank of flotation cells similar to those used in the 1920s at the California Rand. These machines separate particles of different minerals from one another in a water bath, due to differences in mineral grain wetability and their reaction to the surface tension of liquids. Note: The paddles in the center of the picture are blurred, as they are rotating around their horizontal axis. (Allan B. Marquand Collection, San Juan County Historical Society, Colorado.)

Chapter 3
Silver Mining at Red Mountain

Following the decline of gold mining in the district in 1918, silver mining on a major scale began the next year with the discovery of the California Rand. The California Rand, later renamed the Kelly Rand, or simply, the Kelly, became the major silver producer of the district through the 1930s. The mine, in the town of Red Mountain, two miles southeast of Randsburg, worked ore in which silver-antimony sulfides were prominent, with some of the ore carrying considerable gold. Ore of shipping grade was also obtained from the Santa Fe Mine, bordering the Kelly on the northeast, and in the Coyote Mine, bordering the Kelly on the southeast. Several other workings, some quite extensive, were driven in unsuccessful attempts to find extensions of the high-grade silver mineralization. These include the Kelly Rand Extension, Big Four, Silver Bell, Navajo and Swastika, and Flat Tire. The silver mines and nearly all of the silver prospects in the Rand District are restricted to an area about two miles long and 1.25 miles wide (Wright et al., 1953, p. 133).

The discovery of the California Rand (Kelly) was made in April 1919 by Jack Nosser and Hamp Williams who were staking claim corners for J.W. Kelly, sheriff of Kern County. As the men were resting in the shade of a bold rock outcrop, Williams thought he recognized horn silver (cerargyrite, silver chloride) in the outcrop. Samples taken to Kelly and assayed in Bakersfield were found to contain $300 in silver and three ounces[6] of gold per sample (Wright et al., 1953, p. 133). In just the first four years of operation, the California Rand made $7,000,000 (Nadeau, 1965, p. 291).

---

[6] Three ounces of gold in 1919 was worth $62 dollars, as gold was valued at $20.67 an ounce until 1933.

Headframe of Number 2 shaft and mill of the California Rand (Kelly) Mine, Red Mountain Mining District, August 1969. The shaft reached a depth of 1,003 feet and had about seven miles of underground workings. The mine's recorded output of silver is $16 million, making the California Rand the largest single source of silver in California. (Photo by D.D. Trent.)

The silver values were obtained from two sets of veins that cut through the Rand Schist. Bodies of high-grade silver ore were localized in these veins due to nearly perfect closure by a fault known as the "mudwall" on the hanging wall side of a mass of schist that lies within the body of Atolia granodiorite. In addition, there appears to have been repeated reopening of fissures due to ongoing faulting, with continuous mineralization in the same few vein sets. The mine was developed by several shafts that connected with more than seven miles of drifts and crosscuts.[7] Maximum depth of development reached 1,003 feet (Hulin, 1925, p. 110). The silver ore is mainly in the Rand Schist, although several mineralized veins occur in other rocks.

The most abundant silver mineral in the deposit is miargyrite (silver antimony sulfide) with cerargyrite (silver chloride, commonly called

---

[7] In hardrock mining, a drift is a horizontal working that is driven along the trend of a vein; a crosscut is driven at more-or-less right angles across a vein.

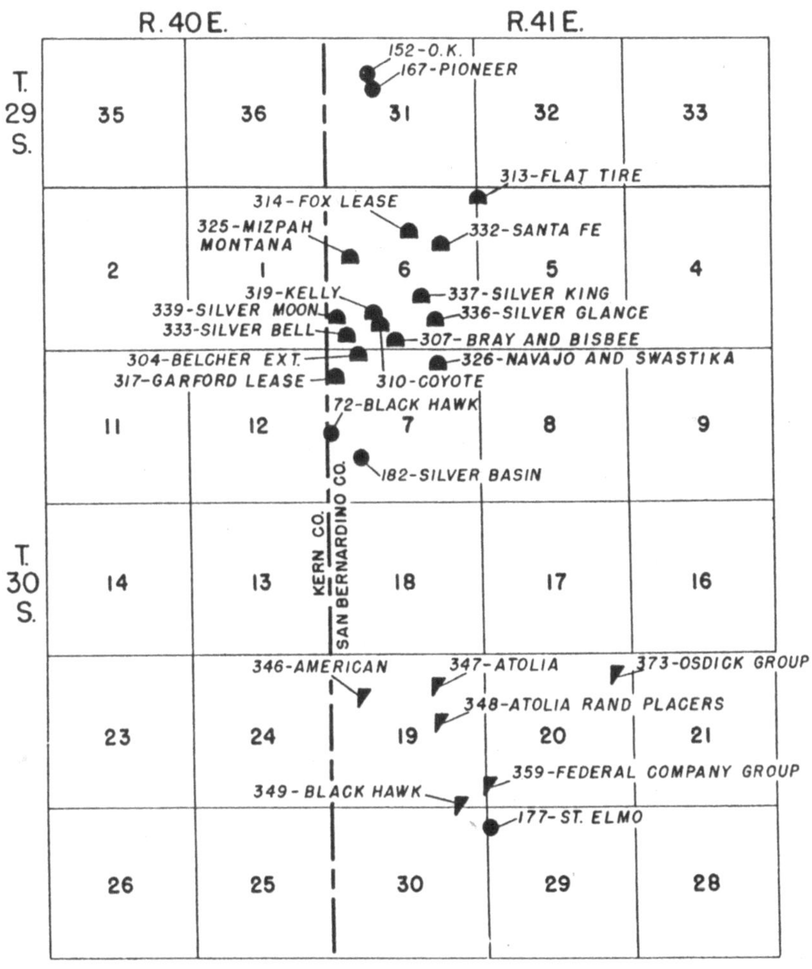

Map of the Red Mountain area of the Rand Mining District, showing the location of the principal silver and tungsten mine workings. (Wright, Stewart, Gay, and Hazenbush, 1953, used with permission, California Geological Survey.)

horn silver) running a close second. Other minerals include pyrargyrite (another silver antimony sulfide), proustite (silver arsenic sulfide, commonly called ruby silver), pyrite (iron disulfide, commonly called fools gold), chalcopyrite (copper iron sulfide), stibnite (antimony sulfide), mariposite (chromium mica), and gold. The gold values were erratic. In general, the ratio of gold to silver was higher in the lower-grade ores than in the rich silver ores, suggesting that the silver and gold had different origins (Hulin, 1925 pp. 97-99).

Despite the expiration of the Pittman Act in 1923, which collapsed the price of silver, the California Rand continued mining silver profitably. By 1926 the mine had grossed over $13,000,000 and paid dividends totaling $4,500,000. However, only three years later, the mine could no longer be operated profitably and was sold for $50,000 (Wynn, 1963, pp. 240-242). The California Rand is noted for pods of unusually high-grade silver ore in relatively small rock masses, some of it assayed as high as 13,000 ounces of silver per ton (Hulin, 1925, p. 99; Fife et al., 1980, p. 546). Eventually the California Rand had a recorded output of $16 million, making the mine the largest single source of silver in California. Except for sporadic development efforts to reopen the mine in the late twentieth century, the California Rand has remained idle for several decades.

Chapter 4
Tungsten Mining in the Atolia "Spud Patch"

Tungsten was discovered in the Randsburg area in 1904 by placer gold miners in the Stringer District and at the Saint Elmo Mine near Atolia. Wynn (1963, p. 228) reports that Atolia's unusual name came about by combining parts of the names of two prominent tungsten miners, Atkins and De Golia, who built the first tungsten mill in the district in 1907. Commercial tungsten production began in 1905 by the Atolia Mining Company, which eventually controlled most mining in the productive part of the district, and produced 95 percent of the tungsten from the district (Lemmon and Dorr, 1940, p. 207; Vredenburgh et al., 1981, pp. 199 and 205). Exploration was stimulated by the demand for tungsten during World War I, with the most extensive development at Atolia between 1914 and 1918, the major producer being the Union Mine.

The bedrock deposits of the tungsten ore mineral, scheelite (calcium tungstate), occur in quartz-carbonate veinlets and in massive chunks in the larger veins in fault zones in tactite, where an irregular pendant of the pre-Cretaceous metasedimentary rocks, including limestone of the Kernville Series, is caught up within and is in contact with quartz diorite which, when intruded, had heated and metamorphosed the limestone (Lemmon and Dorr, 1940, p. 217; Troxel and Morton, 1962, p. 309; Wright et al., 1953, p. 143). Hot mineral-laden fluids streamed off the quartz diorite magma, entered cracks in the limestone, combined chemically with the limestone's calcium, cooled, and crystallized into veins of scheelite.

A cluster of scheelite-rich veins at Atolia occur in an area about two miles long and 500 feet wide. Between 1940 and 1943, one notable vein, 50 feet long and ten feet wide at the Rocky Point Mine, yielded $13,000 in concentrates (Troxel and Morton, 1962, p. 51). Most

Headframe of the New Number 1 shaft (left) and the Old Shaft (right) of the Union Mine, Atolia Mining District. One of the major tungsten producers of the district, the Union Mine was located on the western edge of the productive belt on the property of the Atolia Mining Company. (Hulin, 1925; used with permission, Department of Conservation, California Geological Survey.)

of Atolia's tungsten production was from two other veins, the North and South veins, in the Union Mine of the Atolia Mining Company in the western part of the district. The Union had six shafts and two miles of underground workings. Union Mine's New Number 1 shaft reached a depth of 1,050 feet and had five miles of underground workings. In other mines in the district, most of the shallow ore veins pinch out at depths of only 170 to 260 feet (Fife et al., 1980, p. 544). By 1940 Atolia had 71 shafts with 61,000 feet of underground workings. In addition to the six shafts of the Union Mine, other important mines of the district included the Amity, which was 200 feet deep with 1,100 feet of drifts and crosscuts, and the Attila, 50 feet deep with 110 feet of drifts (Troxel and Morton, 1962, p. 51). In 1943 the company's operations were taken over by the Hoefling Brothers, later known as the Surcease Mining Company (Wright et al., 1953, p. 141).

The Atolia veins also served as the source of tungsten in extensive alluvial placer deposits. These placer deposits were formed in ancient stream channels that lie buried within alluvial fans lying east of the vein deposits, mostly east of today's U.S. Highway 395. Workings in

Drag line surface mining for tungsten, with trommel and conveyor system, in the Atolia District, circa 1945. The mining effort was to locate ancient stream channels buried by the younger alluvium, and then drift along the old channels where the high-density scheelite had been concentrated by running water. (San Bernardino County Museum.)

the old buried stream channels produced gold values as well as tungsten (Troxel and Morton, 1962, p. 51). Most of the placer scheelite ranges in size from fine-grained sand to cobbles that are the shape and size of potatoes, and the scheelite masses could be dug up much as one might dig up potatoes, hence the area became known as the "Spud Patch." A few spectacular boulders of nearly pure scheelite were found that weighed as much as several hundred pounds. The western mines of the district—the Union and Amity—have been about three times as productive as the major eastern mines—the Flatiron, Papoose, Par, and Paradox (Wright et al., 1953, p. 143).

Between 1909 and 1940 the grade of ore ranged from 1 to 15 percent tungstate (WO_3), with an average grade of 4.14 percent. Some of the ore assayed 60 to 70 percent tungstate, an ore grade so rich that it was sacked right in the mine and shipped without milling. Lemmon and Dorr (1940, p. 209) report that the "richness of the ore led to high-grading and to the theft of concentrates, and scheelite 'spuds' were an

accepted medium of exchange in the bars and stores of the region."
The richest stream channel deposits were mined primarily by underground methods. The Spud Patch was penetrated by as many as 400 shafts, ranging in depth from a few feet to about 60 feet, from which drifts[8] were drilled to follow productive buried stream channels. In later years, from 1942 until 1949, Surcease Mining Company dug large open cuts to uncover scheelite-bearing channels (Wright et al., 1953, p. 141).

Because of scheelite's high density, the mineral having a specific gravity of six, it is separated easily from other rock materials in placer deposits by relatively simple gravity sorting processes. A typical concentrating system for recovering tungsten, and gold if present, might include a jaw crusher, a trommel screen, and a wet concentrator such as

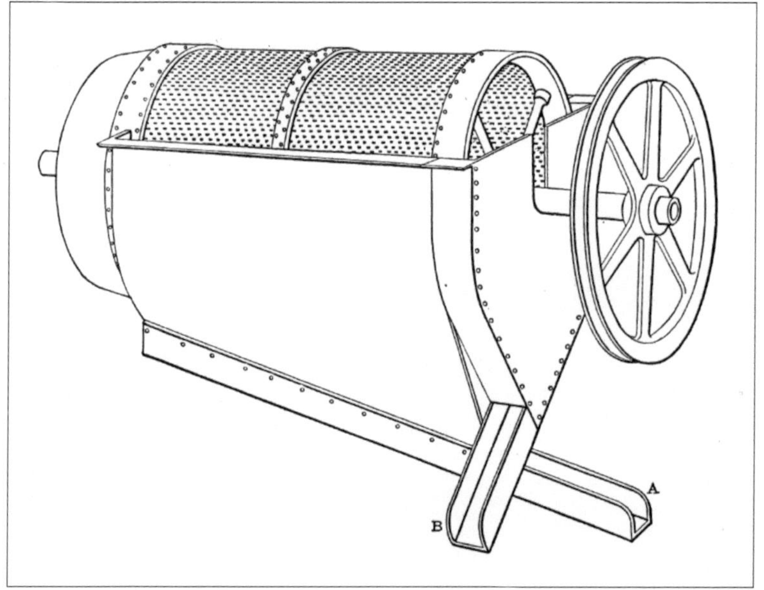

*A trommel, such as used at Atolia in processing scheelite, is a rotating cylindrical screen into which is fed crushed ore, and the undersized and oversized are routed to the discharge spouts **A** and **B**, respectively. (Richards, 1909.)*

---

[8] Drifts in placer deposits are more-or-less horizontal workings that are driven laterally from a shaft and follow the courses of the ancient channels of the buried stream beds.

a Wilfly table. Trommels are revolving cylindrical screens, which may be from six to nine feet long and usually three feet in diameter, used for sizing and gravity concentration. The ore is fed into the trommel, the undersized and oversized materials are separated, and each size fraction is routed to separate discharge spouts for further processing; the oversized return to the jaw crusher for additional crushing, and the undersized to a concentrating table, described earlier, where a concentrate is produced.

Because of the lack of abundant water in the Atolia area, various forms of homemade hand-operated and powered jigs, dry concentrators, and sluices were commonly utilized in recovering the scheelite. Furthermore, because scheelite fluoresces in ultraviolet light, hand sorting under black light was a common practice. The concentrate obtained by these methods was sacked for shipping.

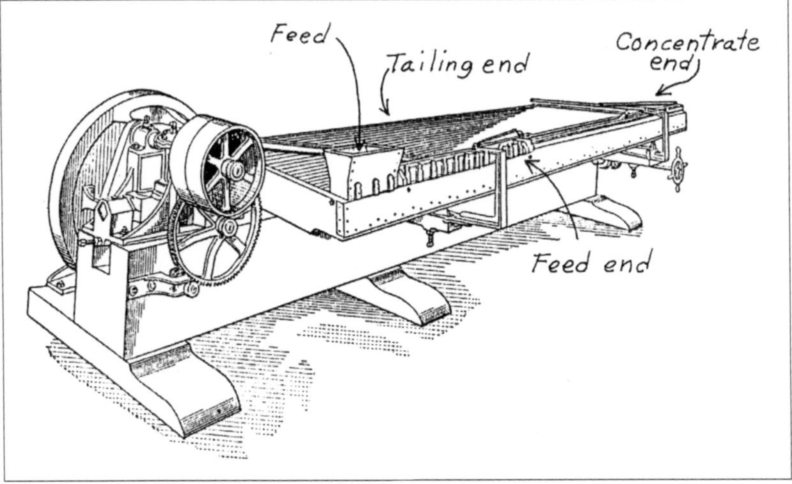

There were many variations of concentrating or jerking tables. They all functioned on the principle that mineral particles can be sorted by density and/or size if kept mobile in water and agitated slightly. One of the more popular brands of concentrating tables, the Wilfly table, is illustrated here. A slurry of sand-and-clay-sized mineral grains and water was introduced at the near side, the "feed side," of the diagram, where the riffles are shorter and the desired concentrate is directed to the "concentrate end" of the table. Riffles are longer on the lower or "tailing side." (Richards, 1909.)

Although production dropped off after 1929, Atolia did not die. The district continued to be the principal source of tungsten in California until 1938 when it was surpassed by production from California's Bishop District (Fife et al., 1980, p. 544). Atolia continued to be a major tungsten producer during World War II and helped make California the nation's largest producer (Needham, 1943, p. 677). Mining activity tapered off following the end of the war, but resumed again for several years in the 1950s, when the United States government began paying $63 per short ton unit[9] of WO_3 for storage in the National Defense Stockpile. Atolia is famous for the largest bodies of high-grade tungsten ore ever to have been found in the U.S. By the mid 1950s, tungsten production for this district totaled more than one million units of WO_3.

---

[9] A tungstate (WO_3) unit is 20 pounds of ore containing 60 percent or more of WO_3. Prior to World War I, a unit of WO_3 sold for $6.50. By 1915 the price reached $14 a unit, and in 1916 it skyrocketed to $33. Following the signing of the Armistice in 1918, the price dropped to $18 a unit, and the major source of tungsten shifted to China, where it could be mined less expensively (Lemmon and Dorr, 1940).

Chapter 5
Modern Gold Mining in the Rand District

The most recent mining activity in the district began in 1984 when the Rand Mining Company, a subsidiary of Glamis Gold, Ltd., obtained rights to work the Yellow Aster and several other historic mine properties on 135 patented mining claims and 530 unpatented lode and placer mining claims. The company began major operations at the sites of historic mines—the Lamont in 1987, the Yellow Aster in 1989, and the Baltic Mine in 1993. The mining targets were large low-grade gold deposits that were best suited to exploitation by open pit and heap leach methods. The proven and probable reserves of the Rand property as of December 31, 2000, were 11,393,000 tons of ore at an average grade of 0.020 Troy ounces per ton,[10] with a predicted yield of 232,700 ounces of gold (Rand Mining Company, 2000). The tonnages involved in

Glamis Rand Mine from Randsburg-Red Mountain Road, September 2002. Leach pad on left, mine offices in center, and waste rock dump on right. (Photo by D.D. Trent.)

---

[10] There are twelve Troy ounces per pound.

mining such a low-grade deposit by open pit, with leach pads and associated waste dumps, are staggering.

Recovering gold from large low-grade deposits by the heap leach cyanide recovery process became widespread in the western United States in the 1980s. Heap leaching is efficient but controversial because of the perceived hazard of using cyanide; also, the large waste rock dumps, open pits, and leach pads that remain, after mining is completed, disrupt the landscape. The process of heap leach gold recovery is illustrated in the diagram on the next page. Ore excavated from an open pit is spread in piles over an impervious high-density polyethylene liner (HDPE) and sprayed with a dilute sodium cyanide solution, commonly about 200 parts of cyanide per million parts of water (a 0.02 percent solution). The solution dissolves gold and silver as it works its way to the bottom of the leach pad, where the leachate solution is caught by a leachate recovery system that directs the solution to the "pregnant" pond from which it is pumped to large vertical tanks at the recovery plant. Activated charcoal in the vertical tanks removes the gold and

Cyanide drip-line system on the surface of the Yellow Aster leach pad, Glamis Rand Mine, April 1993. The dilute cyanide solution, about 200 parts per million (0.02 percent), is kept highly alkaline in order to prohibit the release of hydrocyanic gas. (Photo by D.D. Trent.)

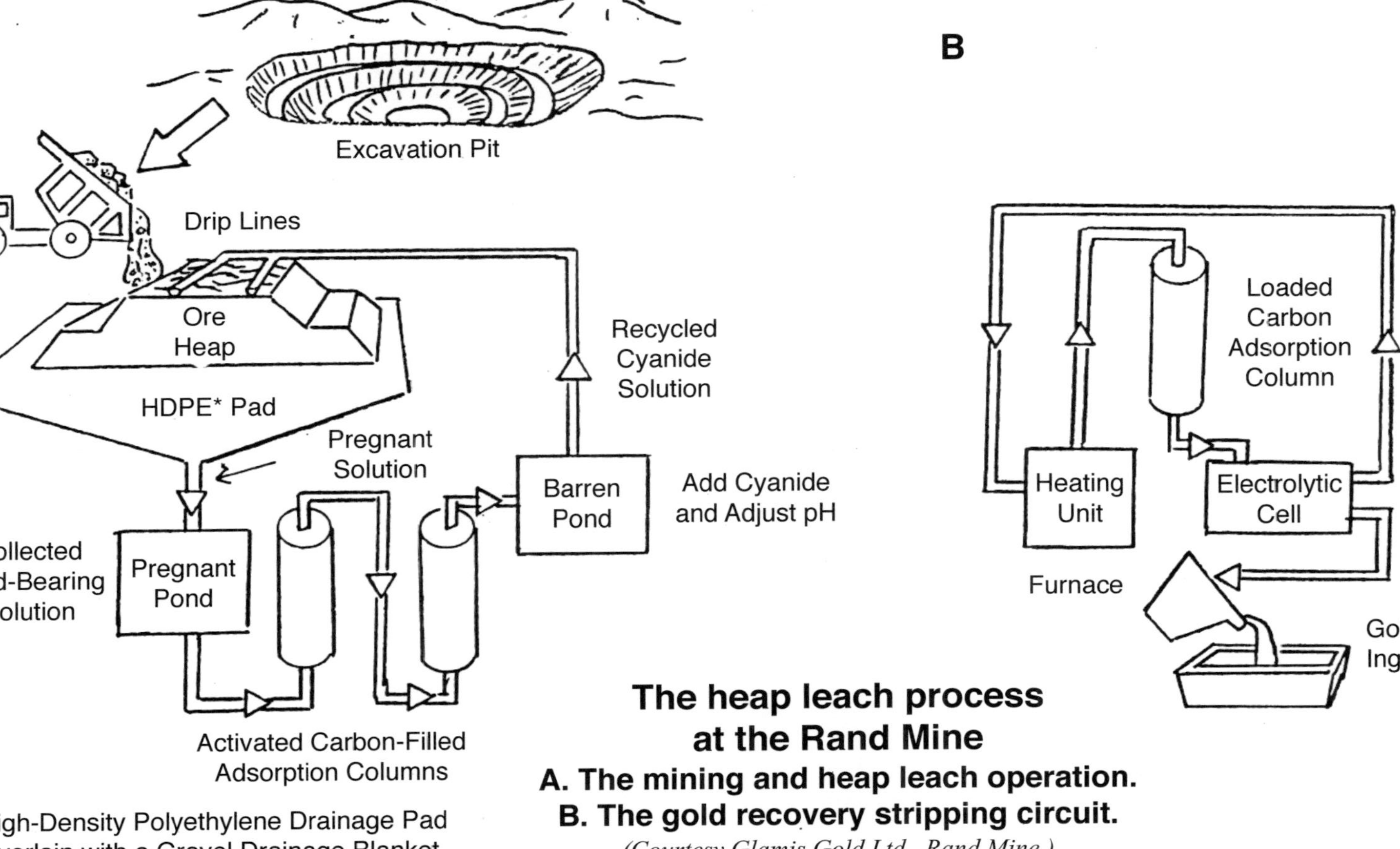

The heap leach process at the Rand Mine
A. The mining and heap leach operation.
B. The gold recovery stripping circuit.
(Courtesy Glamis Gold Ltd., Rand Mine.)

*High-Density Polyethylene Drainage Pad overlain with a Gravel Drainage Blanket

silver by adsorption, and the barren cyanide solution is piped to a mixing pond where it is reconstituted and recycled back to the leach pad. When the carbon becomes loaded with gold and silver, it is treated with a hot caustic solution that strips the precious metals from the charcoal to create a supersaturated caustic-gold solution. This solution is sent to an electrolytic cell that generates a strong ionic (electric) attraction between the precious metals and steel wool, causing the gold and silver to plate out on the steel wool by electrolysis to form a substance called matte. From here the precious metal matte is placed in a furnace where the steel wool produces slag and the precious metals form a melt that is poured into a cone-shaped mold called a doré (Trent, 2005, pp. 366-367). The dorés are then shipped off-site for final refinement to pure gold and silver. A doré from the Rand Mine was usually about 800 fine: 80 percent gold, 10 percent silver, and 10 percent other metals.

Baltic Project gold recovery plant at the Rand Mine, October 2003. The pregnant solution is pumped through a series of treatment tanks (in the center of the photograph) and then passes into the carbon-filled adsorption columns on the left. The barren cyanide solution leaving the adsorption columns is piped to the mixing pond where it is reconstituted and returned to the leach pad. The activated carbon, after being treated in the stripping circuit, is also recycled. (Photo by D.D. Trent.)

Under optimum conditions, the Rand Mine processed an average of 60,000 tons of ore and waste rock per day. A production record was set in 2000, with 100,000 ounces of gold produced at a cost of $176 per ounce. Due to a higher stripping ratio (the ratio of waste rock to ore) in 2001, the production dropped to approximately 75,000 ounces of gold, resulting in an increased production cost of $219 per ounce; and in 2002, gold recovery dropped again, producing only 66,934 ounces. Because of the low commodity price of gold in 2001, plans to develop additional open pits and leach pads were abandoned. Furthermore, the ore bodies already being mined were nearing exhaustion, the Rand Mine's gold ore reserves were approaching marginal values, and the cost of production had increased to about $275 to $320 per ounce. Consequently, the decision was made to shut down the existing operation. Active mining ceased in early 2003, with the large-scale mining equipment transferred to the Glamis Marigold Mine in Nevada. Closure of the Rand Mine began in accordance with the conditions of an approved reclamation plan that requires rinsing the leach pads with fresh water to flush residual cyanide. After rinsing, the pads are drill-tested for

Baltic Project leach pad, cyanide mixing pond, and shipping containers of sodium cyanide, Rand Mine, Randsburg, October 2003. (Photo by D.D. Trent.)

residual chemicals before they are considered reclaimed and the regulating state agencies will allow mine closure. (Glamis Rand Mining Company, 2000 and 2003; Desert Mines, 2003).

Also, in accordance with the site reclamation plan, all equipment and buildings will be removed, and some of the pits will be backfilled, covered with topsoil, and replanted with native vegetation. The plateau-like leach pads, and terraces along the sides of the heaps, will be reconfigured to a natural appearance. Revegetation is done under the direction of professional botanists to assure that the procedure is done correctly. Water bars are placed on the sides of the reshaped leach pads, and sediment collection basins are constructed in order to minimize erosion and prevent storm water from leaving the mine site (Art Champeny, oral comm., 2003). Gold and silver continued to be recovered during the flushing period and by the time the final rinsing ended Glamis recovered almost one million ounces of gold from its Rand Mine operation (Glamis Gold, Rand Mine, 2004). Detoxification, grading, revegetation, puncturing the HDPE liners, sampling of leach pads, and verification were achieved in July 2005, allowing closure of the mine (California Regional Water Quality Control Board, 2005).

References

Barth, A.P., Coleman, D.S., Grove, Marty, Jacobson, C.E., Miller, B.V., and Wooden, J.L., 2003, Geochronology of the Randsburg Granodiorite: Reevaluation of the tectonics of the southern Sierra Nevada and western Mojave Desert; Geological Society of America Abstracts with Programs, vol. 35, no. 4, p. 70.

Brasel, Greg, 2001, Geology of the Randsburg Mining District; unpublished manuscript, Glamis Rand Mining Company.

California Regional Water Quality Control Board, Lahontan Region, 2005, Rescission of waste discharge requirements; Board Order No. R6V-2002-0051 for Glamis Rand Mining Company, Yellow Aster II.

Champeny, Art, 2003, personal interview on October 31.

Clark, William B., 1992, Gold districts of California; California Division of Mines and Geology Bulletin 193.

Cloudman, H.C., Huguenin, Emile, and Merrill, F.J.H., 1919, San Bernardino County: State Mineralogist's Report, Biennial Period 1915-1916; Report XV, part 6; Sacramento, California Mining Bureau, pp. 775-899.

Desert Mines, website accessed January 29, 2003, http://www.desertmines.com/news/012903.html.

Eggers, M.R., and Trent, D.D., 2004, Historic mining equipment and processes within Joshua Tree National Park, in Eggers, M.R., editor, Mining history and geology of Joshua Tree National Park; San Diego Association of Geologists, c/o Sunbelt Publications, P.O. Box 191126, San Diego, California 92159-1126.

Emmons, W.H., 1937, Gold deposits of the world; New York, McGraw Hill.

Fife, Donald L., LaViolette, John W., and Unruh, Mark E., 1980, Gold and mineral wealth of the California desert, field trip guide, in Fife, D.L., and Brown, Art R., editors, Geology and mineral wealth of the California desert; Santa Ana, South Coast Geological Society, pp. 541-553.

Glamis Gold, Rand Mine, California, website accessed December 31, 2000, and January 3, 2004, http://www.glamis.com/properties/california/rand.html.

Grove, Marty, Jacobson, C.E., Barth, A.P., and Vucic, Ana, 2003, Temporal and special trends of Late Cretaceous-early Tertiary underplating of Pelona and related schist beneath southern California and southwestern Arizona, in Johnson, S.E., Paterson, S.R., Fletcher, J.M., Girty, G.H., Kimbrough, D.L., and Martin-Barajas, A., editors, Tectonic evolution of northwestern Mexico and the southwestern USA; Boulder, Colorado, Geological Society of America Special Paper 374, pp. 381-406.

Hess, F.L., 1909, Gold mining in the Randsburg quadrangle, California; U.S. Geological Survey Bulletin 430-I, pp. 23-95.

Hulin, Carlton D., 1925, Geology and ore deposits of the Randsburg
     quadrangle; California Mining Bureau Bulletin 95.
Jacobson, C.E., Grove, Marty, Vucic, Ana, Pedrick, J.N., and Ebert, K.A.,
     in press, Exhumation of the Orocopia Schist and associated rocks of
     southeastern California: Relative roles of erosion, synsubduction,
     tectonic denudation, and middle Cenozoic extension, in Cloos, M.,
     Carlson, W.D., Gilbert, J.G., Liou, J.G., and Sorenson, S.S., editors,
     Convergent margin terranes and associated regions—A tribute to W.G.
     Ernst; Boulder, Colorado, Geological Society of America Special Paper.
Lemmon, D.M., and Dorr, John, II, 1940, Tungsten deposits of the Atolia
     District, San Bernardino and Kern Counties, California; U.S. Geologi-
     cal Survey Bulletin 922-H, pp. 205-245.
Myrick, D.F., 1963, Railroads of Nevada and eastern California, vol. 2, The
     southern roads; Berkeley, California, Howell-North.
Nadeau, Remi, 1965, Ghost towns and mining camps of California; Santa
     Barbara, California, Crest Publishers.
Needham, C.E., 1943, Bureau of Mines Yearbook 1942; Washington, D.C.,
     U.S. Government Printing office.
Rand Mining Company, 1993, Interpretative exhibit at the Rand Mine.
Richards, R.H., 1909, A textbook of ore dressing; New York, McGraw-Hill.
Sagsetter, E.M., and Sagsetter, W.E., 1998, The mining camps speak;
     Denver, Colorado, BenchMark Publishing of Colorado.
Serpico, Phil, 2004, A road to riches: The Randsburg Railway Company;
     Palmdale, California, Omni Publications.
Trent, D.D., 2005, Mineral resources and society, in Pipkin, B.W., Trent,
     D.D., and Hazlett, R.W., Geology and the environment, 4th edition;
     Pacific Grove, California, Brooks/Cole, pp. 364-367.
Troxel, B.W., and Morton, P.K., 1962, Mines and mineral resources of Kern
     County, California; California County Report 1; Sacramento, California
     Division of Mines and Geology.
Tucker, W.B., Sampson, R.J., and Oakeshott, G.B., 1949, Mineral resources
     of Kern County; California Journal of Mines and Geology, vol. 45,
     no. 2, pp. 203-297.
Vredenburgh, L.M., Shumway, G.L., and Hartill, R.D., 1981, Desert fever:
     An overview of mining in the California desert; Canoga Park, Califor-
     nia, Living West Press.
Wright, L.A., Stewart, R.M., Gay, T.E., and Hazenbush, G.C., 1953, Mines
     and mineral deposits of San Bernardino County, California; California
     Journal of Mines and Geology, vol. 49, nos. 1 and 2, pp. 73-147.
Wynn, Marcia R., 1963, Desert bonanza: The story of early Randsburg, Mojave
     Desert mining camp; Glendale, California, Arthur H. Clark Company.
Young, Otis E., Jr., 1970, Western miner; Norman, Oklahoma, University of
     Oklahoma Press.

Index

Modern mining in the Yellow Aster open pit in the 1990s. The haul trucks stand over two stories high and carry loads of about 200 tons. With the mine's average ore grade of 0.020 ounces of gold per ton, an average load of a haul truck yielded only four ounces of gold, enough for eight wedding rings. (Photo by Bill Hample.)